Rainforests
Animal Habitats

Written by Noah Leatherland

©2024
BookLife Publishing Ltd.
King's Lynn, Norfolk
PE30 4LS, UK

All rights reserved.
Printed in India.

A catalogue record for this book is available from the British Library.

ISBN: 978-1-80505-670-6

Written by:
Noah Leatherland

Edited by:
Elise Carraway

Designed by:
Amy Li

All facts, statistics, web addresses and URLs in this book were verified as valid and accurate at time of writing. No responsibility for any changes to external websites or references can be accepted by either the author or publisher.

Photo Credits. Images are courtesy of Shutterstock.com. With thanks to Getty Images, Thinkstock Photo and iStockphoto. Recurring – MirabellePrint, Macrovector. Cover – Macrovector, Hung Chung Chih, GUDKOV ANDREY, KittyVector, dangdumrong. 2–3 – SJ Travel Photo and Video. 4–5 – AustralianCamera, Nicola_K_photos. 6–7 – CherylRamalho, kotoffei, neelsky. 8–9 – Enrico Pescantini, M Gustafson Photography. 10–11 – Tim Fleming, Xico Putini. 12–13 – GUDKOV ANDREY, juan Francisco valbuena, MVshop, tiverylucky. 14–15 – Alamin-Khan, Linas T, Nemanja Cosovic, Stephen Clarke. 16–17 – Abdul Fattah19, Nickolas warner, Petr Muckstein. 18–19 – Jan Hejda, Karel Bartik. 20–21 – Alves-Silva K. R, curiosity, Vaclav Sebek. 22–23 – AustralianCamera, EcoPrint, Lili Kudrili.

Contents

Page 4	What Is a Habitat?
Page 6	Bengal Tiger
Page 8	Sloth
Page 10	Toco Toucan
Page 12	Ring-Tailed Lemur
Page 14	Red-Eyed Tree Frog
Page 16	Goliath Beetle
Page 18	Boa Constrictor
Page 20	Harpy Eagle
Page 22	Animal Habitats
Page 24	Glossary and Index

Words that look like <u>this</u> can be found in the glossary on page 24.

What Is a Habitat?

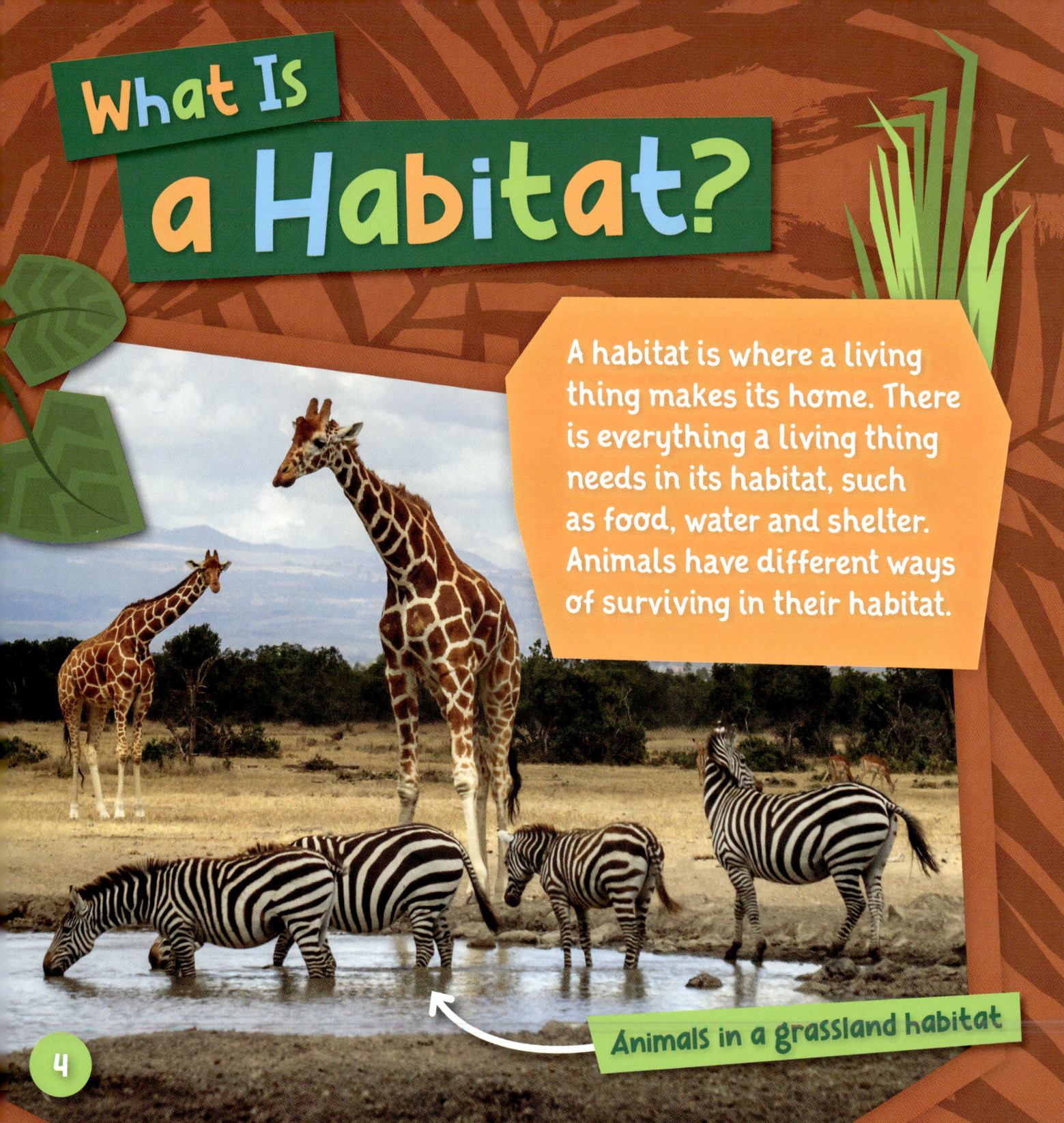

A habitat is where a living thing makes its home. There is everything a living thing needs in its habitat, such as food, water and shelter. Animals have different ways of surviving in their habitat.

Animals in a grassland habitat

Rainforests

Rainforests are forests with lots of trees that get a lot of rainfall. There are a few layers in rainforests. There are lots of amazing animals that live in each layer of the rainforest.

Bengal Tiger

Bengal tigers are top <u>predators</u> in their rainforest habitats. Their stripes help them blend into the bushes, grasses and trees of the rainforest. This helps them stay hidden from their <u>prey</u>.

Many Bengal tigers live in the rainforests of Asia.

When a Bengal tiger catches their prey, they might not eat it all straight away. They use bushes in the rainforest to hide their food so that they can come back to it later.

Sloth

Sloths are found in South and Central America.

Sloths are known for being very slow animals. Surprisingly, this helps them to survive in the rainforest. Moving slowly means that sloths are less likely to be spotted by predators.

Being slow also helps sloths hide in other ways. <u>Algae</u> grows on sloths' fur because they do not move much. This makes their fur look green and keeps them hidden in the trees.

Toco Toucan

The toco toucan is the biggest type of toucan. They have very long <u>bills</u>. Their bills help them keep their bodies cool in the warm rainforests they live in.

Toco toucans live in South America.

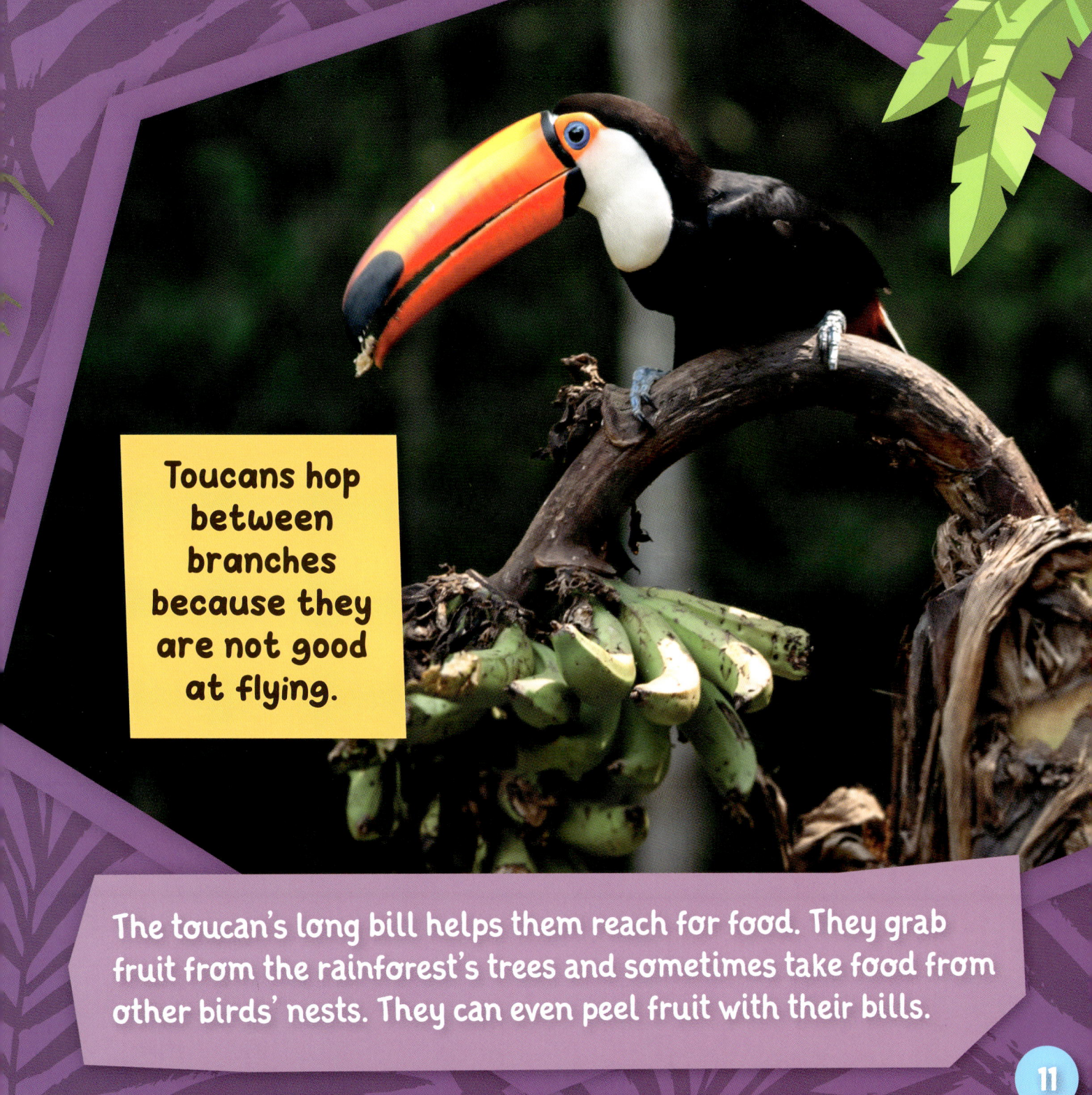

Toucans hop between branches because they are not good at flying.

The toucan's long bill helps them reach for food. They grab fruit from the rainforest's trees and sometimes take food from other birds' nests. They can even peel fruit with their bills.

Ring-Tailed Lemur

Ring-tailed lemurs spend a lot of time on the ground as well as climbing through the rainforest's trees. Their fingers and toes let them grab onto branches. Their tails help them to balance.

A lemur's fingers

Lemurs have strong legs that make them very good at jumping. They can leap from tree to tree and move through the rainforest much quicker than running along the floor.

Lemurs are only found in Madagascar.

Red-Eyed Tree Frog

Red-eyed tree frogs live in the trees. They have suckers on their feet that cling onto the leaves. They lay their eggs on the leaves above ponds. Then, their young fall in when they <u>hatch</u>.

Suckers

Red-eyed tree frogs are found in Central and South America.

Red-eyed tree frogs use their colours to distract predators. Before a predator can attack, they open their bright eyes wide. The flash of colour surprises the predator so that the frog can get away.

Goliath Beetle

Goliath beetles are found in African rainforests.

Goliath beetles are very tough insects that are well-suited to survive the rainforest. They have a strong exoskeleton that protects them. They can also lift things over 800 times heavier than themselves.

Goliath beetles eat sugary things such as tree <u>sap</u> and fruits. They have very strong jaws that can bite through tough plants and the bark of rainforest trees.

Boa Constrictor

The colour and pattern of a boa constrictor's scales help them blend in with the rainforest's plants. This lets them sneak up on their prey. They wrap their bodies around their prey and squeeze.

Boa constrictors are found in Central and South America.

Bushes, leaves and trees can get in the way of a boa constrictor's <u>vision</u>. Instead of finding prey by sight, they use other <u>senses</u>. They can smell their prey and feel their body heat.

Boa constrictors use their tongues to find smells.

Harpy Eagle

Harpy eagles are strong enough to grab their prey from out of trees. They have shorter wings than other eagles. These shorter wings help them fly between branches in the rainforest.

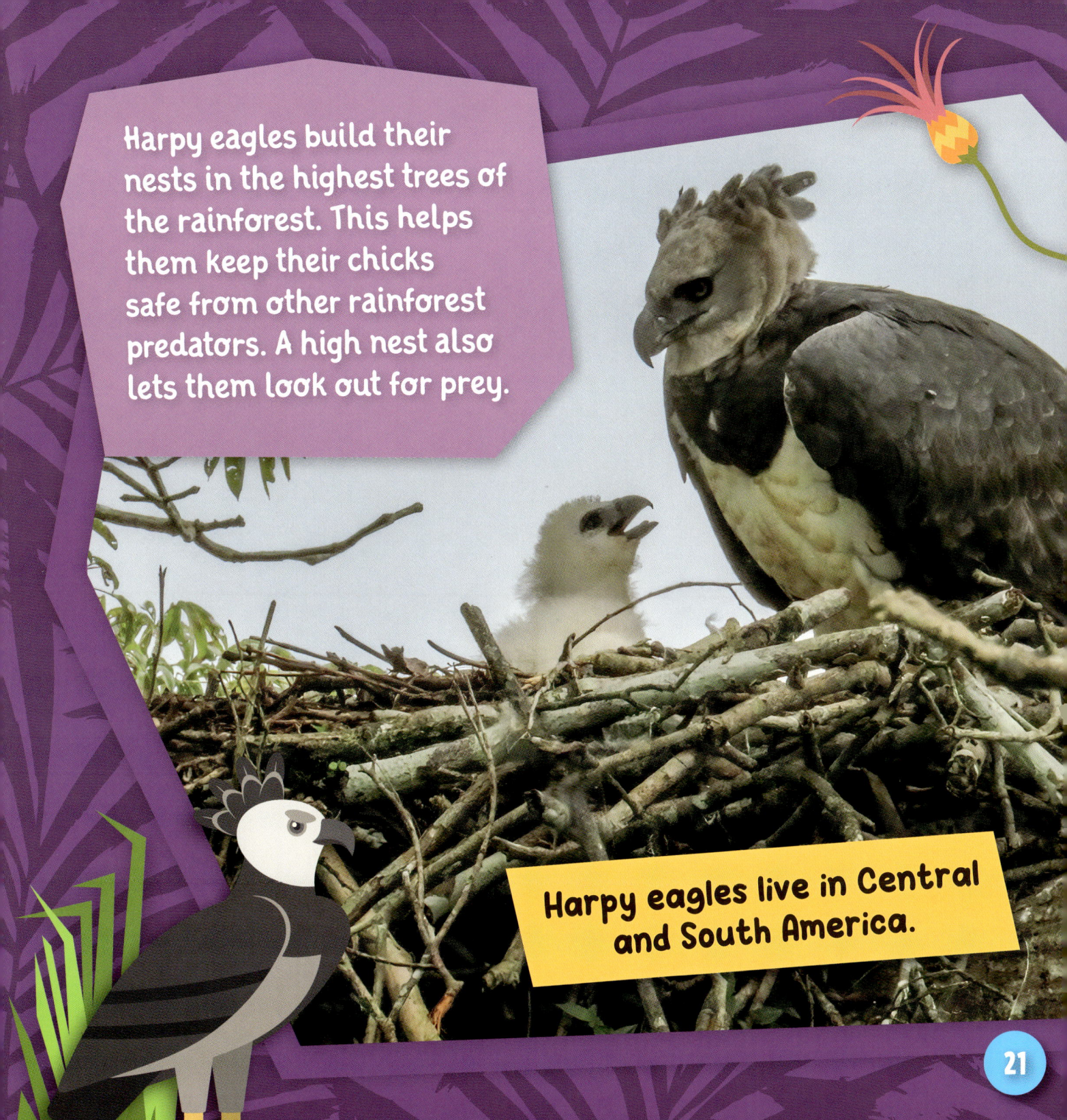

Harpy eagles build their nests in the highest trees of the rainforest. This helps them keep their chicks safe from other rainforest predators. A high nest also lets them look out for prey.

Harpy eagles live in Central and South America.

Animal Habitats

Lots of animals can be found in every part of the rainforest. They might live high in the trees or down low on the ground. They are all important in keeping the rainforest ecosystem healthy.

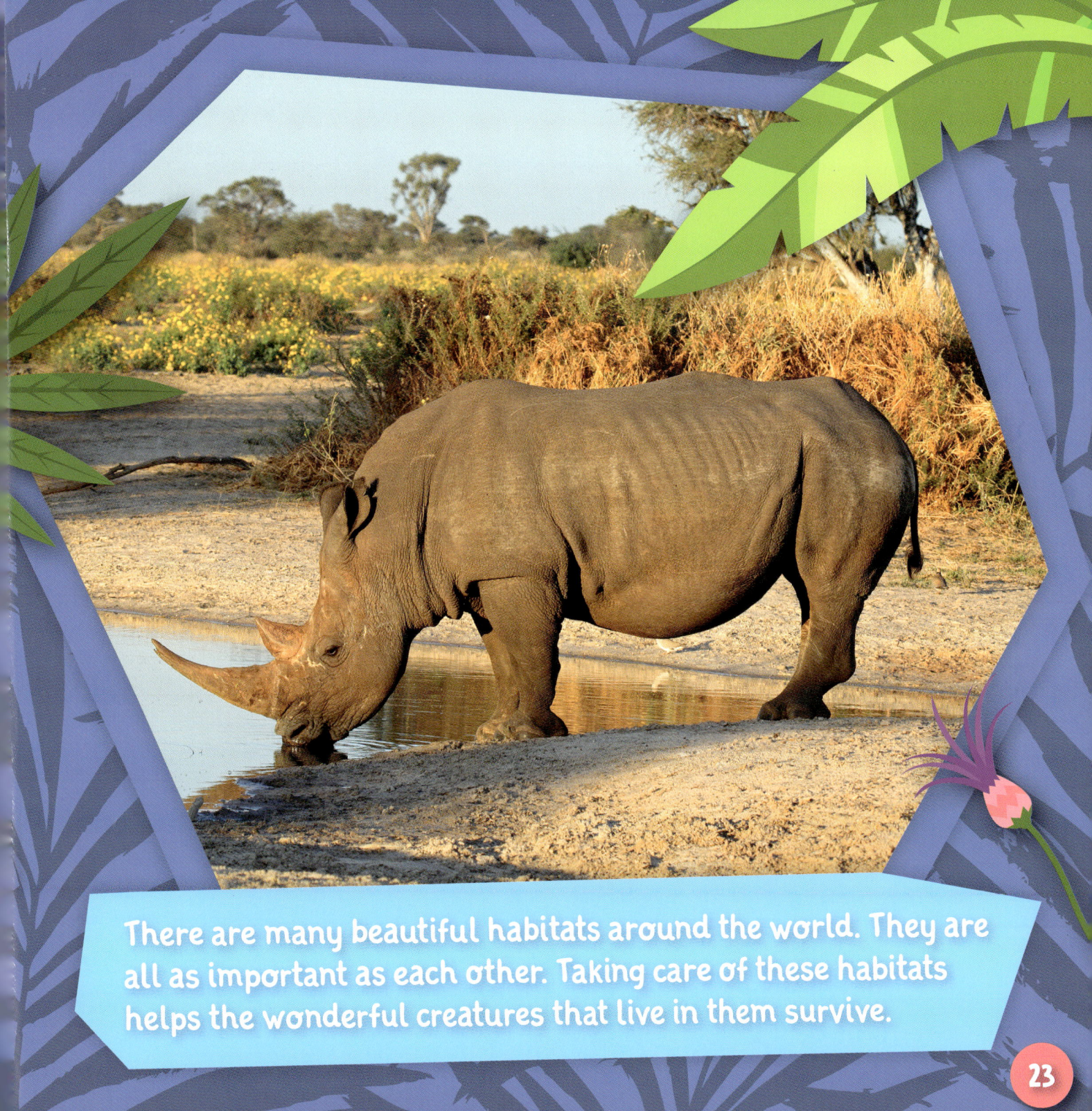

There are many beautiful habitats around the world. They are all as important as each other. Taking care of these habitats helps the wonderful creatures that live in them survive.

Glossary

algae	a plant-like living thing that has no roots, stems, leaves or flowers
bills	another name for the beaks of some types of animal
ecosystem	a community of living things and the environment they live in
exoskeleton	the hard structure on the outside of some creatures
hatch	when a baby animal comes out of its egg
predators	animals that hunt other animals for food
prey	animals that are hunted by other animals for food
sap	a sticky liquid found inside trees and other plants
senses	feelings of things happening around or within a person or animal
vision	sight

Index

algae 9
fruits 11, 17
insects 16
ponds 14
predators 6, 8, 15, 21
scales 18
stripes 6
tails 12
wings 20